The Venetian
A rational progression

Claus Del Mar

Copyright © 2020 Claus Del Mar

Contents

The Venetian i
The Venetian 1
The balances 3
Balances in disharmony 7
Progressions 10
Martingale 11
D'Alembert system 13
Counter D'Alembert 15
Limited counter D'Alembert 17
Dutch 18
The Venetian 21
The prolonged Venetian 25
Repetitions 26
Numerical series 28
Progressions by three 30
Conclusions 32
An evening at roulette 33
Martingale 34
D'Alembert 37
Counter D'Alembert 40
Limited counter D'Alembert 43
Dutch 46
The Venetian 51
Progressions 54

The Venetian

"Mater artium necessitas": necessity is the mother of skill.
Necessity sharpens the wits.
This ancient Latin proverb describes necessity as the force that pushes us to think, to elaborate, to create, to achieve a goal. Often, it has been the determining force at the base of an invention, a discovery. Finding a winning system at the roulette wheel is a challenge to oneself, and the need to win this challenge directs the thought to examine the possibilities of our hypotheses.
A system is a method of play that allows us to play with rationality. The goal of this system is to find a progression of bets to use in order to recover losses without having to resort to huge bests, and, trying to do so in the shortest time possible.
The easiest thing is to build it based on simple chances so that you can only face two possibilities: red or black, even or odd and manque or passe (numbers from 1 to 18 or numbers from 19 to 36). Once elaborated in a satisfactory manner, we will be able to evaluate its use on the other chances.
A progression is a numerical series to recover lost bets, used at roulette in increments or in decrements.
The main characteristic of the Venetian progression is its linear alignment. The sequence of numbers starts from number one and can extend to infinity following a simple principle, without the need for tables or predefined schemes. Conceived for the recovery of lost tokens at roulette, it can obviously be used for other games or situations where you want to recover the invested amount in a short time. The time, considered as the number of sorties, is the main advantage of this progression, as it is halved compared to the number of lost bets.

The Venetian progression, designed to recover the lost capital, can be used to obtain, in addition to the recovery, a win corresponding to the last bet.

A very important argument for the use of any progression concerns the achievement of the balance between the chances.

Another argument to consider is the arrangement of roulette numbers on the green carpet, looking for a percentage of advantage.

The D'Alembert system and some of its many variations, the theoretical waiting time for the repetition of a chance and other numerical series, are themes taken into consideration in this book.

The balances

17 noir, impair, manque. The croupier announces the number, where, the not always loved ball has stopped, creating small or big joys, and other less pleasant emotions.
Now all the roulettes report the last numbers that came out, years ago the players marked them on special cards provided by the gambling house, nothing else was allowed at the table, no sheets, no notebooks, no notes. For many players, the mechanism remains the same: consult the list hoping to find an indication for the next bet. The color, red or black, jumps out at once and an imbalance between the two often provides an indication for a bet. Whoever favors the color that is in frequency, or, who prefers to bet on on the color appearing less often, which is expected to come out for compensation. Often one is led to think that after so many blacks, "of course" a red number will come out, or vice versa; there is nothing more wrong, the numbers, as mathematicians say, have no memory, so everyone has the same probability to come up at any time, either in a perfect balance between the two chances, or after ten, or twenty, or more sorties of the same. The mathematical ones are the only real probabilities.
Three times, for example, we will have the two chances (Red/Black), elevated to the power of 3:

1°	R	R	R	R	N	N	N	N
2°	R	R	N	N	R	R	N	N
3°	R	N	R	N	R	N	R	N

We can notice that after two Blacks (or Reds), at the third time the possibility that a Red number comes out, is exactly 50%.

This is valid for any number of times we consider, after ten blacks, for example, the eleventh sortie will have 50% for black and 50% for red. Whether it is three, ten, a thousand or millions, the possibility remains the same.

There is the belief that after a large number of tests, the two chances tend to balance, for example, if, after ten times there are eight Reds and two Blacks, if keeping on trying, they will equalize or at least get closer. After 100 times I will have about 50 and 50, after a thousand tests about 500 and 500 and the more you increase the number of tests, the closer the balance between the two chances.

If we analyze the possibilities in a mathematical way, this is not true. Starting from eight blacks and two reds, we add the mathematical sorties for 10 that is 1024 strips of 10 numbers each.

Starting with 10 blacks we will have 18 blacks and 2 reds, with 10 reds, the blacks will remain at 8 and the reds will rise to 12. This, however, only once in 1024. The highest frequency is the combination of 5 reds and 5 blacks with 252 presences, followed by 6 blacks and 4 reds and 6 reds and 4 blacks with 210 each. Adding to our base combination (8 reds and 2 blacks) we will have 252 combinations with 13 blacks and 7 reds, 210 with 14 blacks and 6 reds and 210 combinations with 12 blacks and 8 reds. Only 45 combinations will allow us to achieve the perfect balance: 10 blacks and 10 reds, the combinations with 2 blacks and 8 reds. With the 120 combinations of 3 blacks and 7 reds, we will approach with 11 blacks and 9 reds.

Achieving balance is a transitory situation, which can occur, quite randomly, in a short time or a long time afterwards.

If we only do one test by pointing one of the two chances, red or black, 20 times in a row, the probability that a combination with 9, 10 or 11 Red (or Black) is about 50%, increasing the combinations to 8 and 12, the probability rises to about 75%. So in the case of one or a few tests, this percentage remains in our favor.

By increasing the number of tests, the probability tends to go down to 50%. From all this, we can deduce that the probability of a red (or black) number coming out and the number of times we will spot it are inextricably linked.

Several years ago, I went to a casino with a friend, convinced that if he could cover the lost bets up to thirty times, he would certainly always win. In the car, on the way back, we would inevitably talk about our bets, about the "bad luck" that sometimes seems to be going against a player, and about the "luck" that another player had had at our table.

Although I followed his reasoning carefully and listened to his motivations just as carefully, based on events reported in books and newspapers, showing that the very long sequence of a color was a very rare event and should not have been repeated more than a certain number of times, I did not feel I could exclude the possibility of it happening.

If we develop a series of 30 trials, considering only the two possibilities, red or black, we can notice that we will have 1.073.741.824 sequences and only one of them completely red and one black.

These are enormously large numbers and they mean that we would have to go to the table, to expect that event, over 61,200 years if every play would take only a minute. So, the fact that this will happen in the lifetime of an avid player is an extremely small possibility, but we cannot exclude it. In fact, there is nothing to prevent it from happening on the first day of the game and on the other hand it is not guaranteed that it will happen in 61,200 years, it may be delayed.

*

In the evenings spent at the roulette tables, I saw players relying on luck, those who played their lucky numbers, those drawn from dates dear to them, those who followed the inspiration of the moment until some other player who relied on the random toss of the token on the green carpet. There were those who, instead, scrupulously followed systems devised by them or learnt in some book. The study of the throw of the ball, the "hand" of the croupier, was also a frequent topic among players, according to which the croupier would achieve an automatic throw, often keeping the speed of the ball constant. From this, one could predict an area, where the ball will fall.

*

From cold and logical mathematics we know that numbers have no memory, it follows that my bets will not necessarily be at the same table, one after the other. I can take a break and resume at another table, I can leave and resume the next day, and I could even place my bets changing tables every tome, or leave the game today to resume it in a month.

This is true, even if we play with systems, for example, if I apply a progression to my bets and I find myself at -8, I don't need to wait at that table to go up to 1, I can continue another time and at another table. The result does not change, I could win, or not, continuing on the same table or resuming the game at any other time.

Balances in disharmony

The 36 roulette numbers, zero aside, are in perfect balance. There are 18 red numbers and 18 black numbers, 18 even numbers and 18 odd numbers, manque and pass numbers or the first 18 numbers and the other 18 larger numbers. So balance reigns supreme, but a careful player will not have missed their arrangement on the green carpet. The normal alternation of color stops at the number 11, which is black like 10, then the alternation resumes until the number 18, at that point the 19 is not black as you would expect, but red, the same thing again for the numbers 28 and 29 both black.

At the origin of the game, the colors were assigned according to a principle: the sum of the digits. To the even sums the blacks, and to the odd sums the reds. For example 17 is equal to 8 (1+7), even for black numbers and odd for red. So 17 is a black number while the number 32 is red (3+2=5).

From this disharmony, it results that red numbers are not divided in balance between even and odd, in fact, there are ten red and odd numbers and eight red and even numbers. Reverse situation for black numbers: ten even and eight odd. That implies that if I bet one token on red and one on odd, in the case of a higher frequency of red numbers, I would have a higher win.

In the three columns only in the first one the red numbers are six like the black ones, in the second one there are eight black and four red, in the third eight red and four black. The imbalance continues in the six line, the square and the double street.

What advantage can this disharmony give the player? I, if I knew that, I would have skipped this subject..., perhaps, some of the readers will have a more valid idea for a profitable system.

In recent years, I have not found an application that could continue to give a win in the long run, without arriving at too big stakes or too long time. Surely there is that if only red numbers came out, it would be enough to bet on the third column, constantly, even just a token, to have a profit. Just as the second column would give it for black numbers. The red numbers in the other two columns are 10, so I would have a loss of 10 tokens, while the third column would give me a win of 16 tokens, I just need to keep a constant bet on the column to increase, inevitably, the winnings over time. The fact remains that the opposite color is there, so how to do so that its presence does not affect the bet on the column? Between red and black there is balance, so if I bet a token on black, sooner or later, excluding zero, I would find myself with no loss, but how to tie the column bet with the color? I could increase the bet on the column in relation to the opposite color, but even in this case you would have very large bets.

Trials, new ideas and other evidence, but how to take advantage of this disharmony, still remains an unresolved problem.

As already noted, odd red numbers are more than even red numbers. Imbalance also, of course, for even blacks, greater than odd blacks. If the balance were perfect, there should be nine odd red numbers and nine even red numbers, as well as nine even blacks and nine odd blacks. Therefore, by betting two chances, e.g. red and odd, we will have the simultaneous winning of both, once every four bets. Actually, this is not the case. The actual frequency for their combined sortie, (red and odd, black and even) is once every 3.6.

The column and color combination, occurs every six times only if we match the first column to one of the two colors. Red color and third column, as black color and second column, will occur every 4.5 times.

If we consider a series of six numbers in balance, we will have three red numbers and twice the numbers in the third column.

If we place a token on the red, we will have won three times and lost as many. Same thing for the column, two won and four lost.

So in this period of time our expectation could be the simultaneous winning of the red and the third column. To get it, we could chase it by applying an progression to our bets, intended to cover the tokens lost in the inevitable delay.

Progressions

By betting a token, constantly, on the same chance, sooner or later we will win. We could win at the first shot or after many, really many shots, however sooner or later we will win, it's just a matter of waiting for our chance to have a higher frequency than its opposite. For example it could take eleven shots if up to the ninth the opposite chance had been in advantage then, at the tenth we would have reached a tie and finally at the eleventh our chance would have passed to a greater presence from the beginning of our bet (six to five).

This can happen after ten or after a hundred or a thousand or more shots, when it happens, only the "chance" knows, we can only wait for it to happen, and try, with coordinated betting systems, to shorten this time, or, take advantage of it to get more winnings.

Famous people like D'Alembert, have devised orderly and logical successions that increasing in their series, take the name of progressions.

Martingale

The simple progression is that with each lost bet you double the next bet. By initially betting a token, the second bet will be of two tokens, the third of four, the fourth of eight and ahead of the doubling for the following ones.

With this progression you only need to win once to recover all the lost chips and win one. Certainly the most advantageous from a mathematical point of view, not so for other factors.

The first one concerns the potentially huge number of chips used, for example after nine negative bets we would have already lost 255 chips, after twelve even 2047. Betting the next 2048 tokens. What is not possible, this is the second negative factor, because at each table there are limits not to be exceeded with a single bet.

This system, however, indicates a possible progression to be modified making it more affordable.

We can use it only initially when the bet is still minimal, and then move to other systems or continue with its simple principle but in a less expensive way.

There are many possible variants to the martingale, however, all the changes made to the original progression provide that the recovery of lost tokens is done with more winning bets.

For example we can double the progression, repeating twice the bet, both negative and winning.

Original: 1 - 2 - 4 - 8 - 16 - 32 ....
Double: 1,1 - 2,2 - 4,4 - 8,8 - 16,16 - 32,32 ...
Or triple it: 1,1,1 - 2,2,2 - 4,4,4 - 8,8,8 - 16,16,16 - 32,32 ...

In these cases we will have to get two or three winnings to close the progression and return to betting 1 token.

In the chapter dedicated to the progression trials, 120 game sessions, of an evening at the Monte Carlo casino back in 1930, have been taken into consideration. These were used to compare the systems and to determine their yield.

Applying the Martingale, at the end of the 120 rounds, always focusing on the color Red, we would have a profit of 53 tokens. The maximum bet would have been 32 tokens, to overcome a Black sequence of five times.

Obviously this is a lucky case, if for example the sequence of Black numbers would have continued we would have to bet, the next shot, 64 tokens. One more Black and the bet would rise to 128, then to 256 ... Bearing in mind that even if we had the possibility of such high bets, we will go to the threshold of the table. In that evening, we would have had to face twice a negative sequence of five black numbers, but despite the increase in the bet we would never have been forced to put in play other tokens than the first one.

D'Alembert system

Jean Baptiste D'Alembert, who lived in the eighteenth century, has been the creator of important mathematical theorems, physical principles, philosophical treatises, and has contributed to the study of astronomy. Of all this, his fame, still current, is mainly due to its system applied to roulette...

The D'Alembert progression is simple in concept and application: losing goes up, winning goes down.

Starting with a token and losing, the next bet will be of two, the next of three, then of four and so on until the win, then we go down to the previous bet, for example from four we go to three, then if we lose, we go back to four, if we win to two and then to one. When we return to the balance between the two chances our winnings will be equivalent to the number of times we have lost.

For example if we bet on Red in this sequence of sorties:

N	N	R	N	R	N	N	N	R	N	R	R	R	R
1	2	3	2	3	2	3	4	5	4	5	4	3	2

To return to a token bet we will have:
Lost with the Blacks: 18
Won with the Reds: 25
The Black numbers are 7, the Red 7.
The winnings are 7 tokens that correspond to the number of bets lost with the Black numbers.
The completion of this progression is when you return to the initial bet of a token, to get this, the number of the Reds must be equal to the number of the Blacks.

In the testing chapter of the progressions, by applying the progressions to the game sessions taken into consideration, this one, D'Alembert, goes through a negative phase and the results do not give it the honor it deserves.

After a first positive balance that increases the winnings to ten tokens, begins a negative sequence that brings the balance to 169 tokens in negative. A recovery ends the evening with a negative balance of 38 tokens. At this point, we should have continued our game, or, resume it again another night, to complete the progression. We cannot know how long it would have taken to complete it, nor what the maximum bet would have been. From this test, however, we can see the advantage that the D'Alembert progression provides. For example, in the color alternations, we can see that the balance gains one token each time.

Considering that in the test night, the black numbers came out for 60 times and the red numbers for 56 with 4 other red numbers we should reach a profit. In fact, if they came out one after another, our bets would go down, 18, 17, 16 etc. for a final balance of 12 won tokens. This comes to color parity, with the last bet of 11 tokens. Continuing the progression until we return to bet a single token, our winnings would be much higher. Still continuing the hypothesis of a succession of red numbers, we would win 10+9+8+7+6+5+4+3+2 for a total of 66 tokens.

Counter D'Alembert

There are many variations to the D'Alembert progression, one, defined against D'Alembert as it is played in opposite to winning.
While in the original progression, by losing, you increase the bet, in this variant you decrease it, by winning you increase it:

N	N	R	N	R	N	N	R	N	R	R	R	R
1	1	1	2	1	2	1	1	1	2	1	2	3	4

Considering the previous series, as before we will bet on Red, which will give us a win of 13 tokens.
Lost with the Black 10 tokens.
Our winnings will be three tokens.
The advantage of this progression is the reduction of the number of tokens bet especially in the big delays of our chance, the Red in this case, where we will have had to increase the bet, we will decrease it to a single token.
The disadvantage is when there is an alternation of sorties:

N	R	N	R	N	R	N	R	N	R
1	1	2	1	2	1	2	1	2	1

    Black Numbers: 5
    Red Numbers: 5
    Lost Tokens: 9
    Tokens won: 5

Despite the balance between the two colors we will have lost 4 tokens.

In the last chapter, dedicated to the progressions tests, the evening ends with a negative balance, going through alternate positive phases. The first of these leads us to win 17 tokens, and then continue in negative until we return to winning 2. The progression continues to give a negative balance until the end of the evening. In reality we could have continued until we reached a positive balance. However, in this test, we reached a negative balance of 32 tokens, while the traditional D'Alembert progression closed with a balance of 38 tokens, always in negative. The Counter D'Alembert progression proved to be advantageous for the number of tokens to bet.

Limited counter D'Alembert

In the following diagram, always in the chapter about the tests, a limit is applied to the counter D'Alembert progression. When a positive balance is reached, it is interrupted and restarted by betting a token. So in the first bet that we win a token, we don't increase but we bet agaign one. Applying the progression in this way, we reach the positive balance, twice with winnings from one token, and twice with six tokens. Each time, interrupting the progression. For five times, instead, the winnings will be immediate, at the first bet from a token. In the final phase, the trial closes, interrupting the increase of the bet, with a negative balance of 1 token. The winnings obtained correspond to 19 tokens. Also in this case, the progression should be completed to achieve its purpose, i.e. a positive balance, but in order to make a comparison between the progressions, we have considered the same permanence.

Dutch

Another progression very well known by the players is the Dutch. The base remains always the D'Alembert progression and even with this variant we try to reduce the number of tokens to bet. What remains constant in all these variations is the balance between the number of lost and won bets.
The basis of the Dutch progression is the note of the lost bets to decrease with each win by eliminating the lowest one:

R/N	Bet	Win	Balance	Sequence
N	1		-1	1
N	2		-3	12
N	2		-5	122
R	2	2	-3	22
N	3		-6	223
N	3		-9	2233
R	3	3	-6	233
N	3		-9	2333
N	3		-12	23333
N	3		-15	233333
R	3	3	-12	33333
R	4	4	-8	3333
R	4	4	-4	333
R	4	4	0	33
R	4	4	+4	3
R	4	4	+8	

You increase the bet when you cancel all the lost bets at the lower level: when we bet 3 tokens, and we win, we would recover all the 2 tokens bets, then, the new win will increase our bet by one token. From 3 we go on to bet 4 until we recover all the bets of 3 lost tokens.

Note that the progression to follow, starts after losing the second bet, the 2 tokens bet, because if we lose the first bet of 1 token, we immediately move on to bet 2 tokens.

In the evidence table, in the last chapter, we discover that even with this progression we close with a negative balance. When the numbers sector for the trials was decided, it was chosen to always bet on red, if we had bet on black, the progressions applied so far would have closed with a winning balance. Our goal, however, only in these tests, is not to get a win, but to compare the progressions to find out where, and what, are the advantages of one compared to the others. The Dutch progression, first of all, needs less money, in the test we reach, at most, the 8 tokens bet. Its use is quite simple, although it is essential to note down the sequences of bets.

In the first twenty-five bets we always have a positive balance, which reaches its peak at 10 tokens. After that, the balance goes down to the end of the test, to reach the 52 lost tokens. As we decided before, we take into consideration only a small part of the numbers of that evening, continuing, the progression would have reached a positive balance. To be exact, continuing to play on that night, we would have had to wait 97 more shots to delete all the sequences. Obtaining a positive balance of 49 tokens.

Other known variants, looking for less consistent bets, modify the concept of going up, or down, with the bet in an immediate way, as in the case of another progression:

When you lose you continue with the same bet and only go up when you win.

For another, the progression remains the original one, losing you go up and winning you go down, but, you start with an initial bet of 10 tokens, you add or remove a token until you win.

Almost always in these progressions based on D'Alembert, you get a win when you reach the balance between the two chances.

With the Paroli, of which there are many variations, you bet the winnings obtained. It is not a progression whose end is the recovery of the lost tokens, but the loss, always remaining with only one token can be the advantage of our Paroli.

For example, if at the time of winning our balance is negative by 5 tokens, we will establish three rounds for the recovery:

Winning the first of 1 token, then 2 tokens in the second and 4 in the third. By winning all three bets we will have won 7 tokens which will allow us to return to not having a negative balance.

This is just an example of the use of Paroli, everyone can foresee their own, both to return with losses and for a profitable game.

The Venetian

Assuming that the balance between two chances occurs if the number of their sorties is equal, when they are in favor of one of the two chances, the difference between the two represents the imbalance: for example with 26 sorties of which 16 black and 10 red, the imbalance in favor of black is 6.
The Venetian progression allows the recovery of the unbalance with half of the sorties in favor. In the example we recover the 6 blacks with 3 reds.

*

In the normal sequence of the numerical series (1,2,3,4,5...) the number multiplied by the previous number is double the sum of the previous numbers:

1 2 3 4 (5 x 4) = 20 (1+1+2+2+3+3+4+4) = 20

This is the basic concept for the Venetian progression.
Its simplicity and ease of use are evident: bet the same number of tokens, as many times as half the number of previous bets lost.
This is the basic principle of this progression and can be applied following a normal progression: 1, 2, 3, 4, 5, 6 ... Our winning bet, will have to be repeated until we get the number of winnings needed to cover the losses, that is 50% of the times we lost.
The progression applied to the simple chance: for ease of use, the bet, in case of loss, will be double. In the case, hopefully, of winning, the bet will be repeated with equal number of chips for the number of times that corresponds to half of the number of bets lost up to that moment which is nothing more than the previous number in the progression.

An example will clarify the concept:

First bet	Second bet	Third bet
1-	2-	3+
1-	2-	3+

Lost twice the first bet (1) we will increase by one, lost also the two bets of 2 and winning the one of 3, we will repeat it and make up for the losses of four bets with only two won.

By won bets, as well as lost bets we mean the number of bets for that number of chips.

For example:

	Number of bets
1-	-1
1+	0
1-	-1
1-	-2

Once the two lost bets have been reached, at this point we move on to the next bet by two tokens:

	Number of bets
2-	-1
2-	-2

Now the bet is three tokens:

	Number of bets
3+	+1
3-	0
3-	-1
3+	0
3-	-1
3+	0
3+	+1
3+	+2

Let's continue with other examples:

1-	2-	3-	4+
1-	2-	3-	4+
			4+

In this case lost the first bets for a total of twelve tokens (1+1+2+2+3+3), we will continue the four token bet until we get the three winnings, the twelve tokens, or until we lose the four tokens twice and in this case there will be four necessary bets of five:

(5x4) = 20   (1+1+2+2+3+3+4+4)=20

1-	2-	3-	4-	5+
1-	2-	3-	4-	5+
				5+
				5+

By losing you repeat the bet twice, winning, you repeat for half the number of bets.
A further example:

1-	2-	3-	4-	5-	6-	7-	8+
1-	2-	3-	4-	5-	6-	7-	8+
							8+
							8+
							8+
							8+
							8+

As you can see the eight token bets will have to give a positive balance of seven winnings, which will compensate for the fourteen lost bets. 56 will be the lost tokens and the same recovered with seven bets.

Summing up, for the simple chance the number of tokens to bet, in case of loss will be repeated twice, in case of winning it will be repeated as many times as the previous number.

In the previous examples, we have repeated the 3 token bet twice,
3 times the 4 token bet, 4 times the 5 tokens etc. in the last 7 times the 8 tokens.

In the tests, in the last chapter, we reach a maximum bet of 8 tokens, for a balance of 14 in favor of black numbers. Equalized with 7 red numbers, we close the progression returning to the profit of 7 tokens. We continue the evening with another major sequence for 10 black numbers, which we balance with the 5 bets of 6 tokens. We close with a profit of 9 tokens won.

The prolonged Venetian

By doubling the number of bets we can extend the bets on simple chances. This will allow us to place less substantial bets.

1-	2-	3+
1-	2-	3+
1-	2-	3+
1-	2-	3+

After four lost bets of one token and four of two tokens, with four won of three tokens we will reach the draw.
The ratio between lost and won always remains two to one.
The progression can also be extended by tripling, quadrupling ...
Tripled:

1-	2-	3-	4+
1-	2-	3-	4+
1-	2-	3-	4+
1-	2-	3-	4+
1-	2-	3-	4+
1-	2-	3-	4+
			4+
			4+
			4+

Eighteen lost bets, corresponding to thirty-six tokens, are recovered from the nine bets of 4 tokens, won.

Repetitions

I bet to the first six line, a number of the third comes out, I lose. I repeat the first six line and add a bet on the third. Needless to say, another one comes out: the fourth. Then I bet again the first and the third with the addition of the fourth. Once again I lose, a number from the second six line comes out. But every how many trials does a six line repeat? A dozen? A full one? Mathematics, or rather mathematicians, provide us with more or less complex formulas.

A simplified formula:
$$\frac{K*(K-1)}{2n}$$

n= total number of events
K= number that multiplied by itself-1, is equivalent to twice the total number of events. It determines the events necessary for a repetition.
Apply this formula to determine the repetition of a number out of thirty-six:
n=36
To find K we will have to identify the number that multiplied by itself-1 gives the double of n, so in this case it will have to correspond to 72, and exactly 9, that multiplied by itself -1, that is 8, gives the double of the events (36).

$$\frac{9*(9-1)}{2*(36)} \quad \frac{9*8}{72} \quad \frac{72}{72}$$

So given 36 numbers, 9 are (K) the number of events needed for a repetition.

Note that to reach 9, we will have bet exactly 36 times (1+2+3+4+5+6+7+8) =36, which corresponds to the total number of possible events.
ex: repetition sorties

	sorties	repetition:							
1°	35								
2°	4	35							
3°	6	35	4						
4°	21	35	4	6					
5°	16	35	4	6	21				
6°	33	35	4	6	21	16			
7°	10	35	4	6	21	16	33		
8°	12	35	4	6	21	16	33	10	
9°	6	35	4	6	21	16	33	10	12

One more example with six lines:

$$\frac{4*(4-1)}{(6*2)}$$

$$\frac{4*3}{12}$$

$$\frac{12}{12}$$

Where 12 corresponds to twice the total number of six lines and 4 to the number of events needed for a repetition.
Of course this formula represents the ideal average for a repetition, to this we will find both upper and lower sorties.

Numerical series

The normal sequence of numbers (1,2,3,4,5,6...), apparently without secrets, offers us a new ratio by increasing the same series.
x, x+(2), x+(3), x+(4), x+(5), x+(6)...
where x corresponds to the previous sum.

	X	increment
1°	1	
2°	3	(1+2)
3°	6	(3+3)
4°	10	(6+6)
5°	15	(10+5)
6°	21	(15+6)

This ratio of three to one can be used by repeating the bet three times in case of loss and a number of times corresponding to one third of the number of bets lost to break even.

lost	won
1	3
1	
1	

lost	lost	won
1	3	6
1	3	6
1	3	

lost	lost	lost	Won
1	3	6	10
1	3	6	10
1	3	6	10

lost	lost	lost	lost	Won
1	3	6	10	15
1	3	6	10	15
1	3	6	10	15
				15

As you can easily deduce the number of winning bets after six losing ones must be two (Wins 6, 6).

For the recovery of the nine lost bets, we will have to have three winnings (Wins 10, 10)

Four the necessary winnings (Wins 15, 15, 15) to balance the twelve lost. Sixty tokens lost in 12 rounds and 60 recovered with 4 rounds.

Progressions by three

Considering the normal numerical sequence (1,2,3,4, ...) we consider a segment starting from number 1 for any length. The sum of the numbers of this segment is the third part of the segment of equal length starting from the next number.

segment	next	segment
1	2	3
12	3	45
123	4	567

1 is the third part of 3, in the second segment the sum is 3 (1+2) and is the third part of the sum of the next segment 9 (4+5).
In the third example the sum of the first segment (1+2+3) is 6 and the sum of the next segment (5+6+7) is 18.
A further example with the initial segment length 6:

segment	next	segment
1 2 3 4 5 6	7	8 9 10 11 12 13

The sum of the first segment is (1+2+3+4+5+6) =21 which is the third part of the sum of the second segment (8+9+10+11+12+13) =63.
We could use this sequence for a betting system that allows us to recover losses with one third of the bets.

Repeat the losing bets three times until the first winning one that would represent the next and one bet for each number of the following segment for the winnings.

In case of losing one of the bets of the next segment, we bet until we get three losses or one wins. From the three losses we recalculate the new segment, with only those lost by three and start again.

Example:

1 2 3 4 lost three times
5 won
6 won - start of recovery segment
7 lost for three times
The new segment to recover will be 5 numbers long.
8 won
9 won - start of recovery segment
10 won
11 won
12 won
13 won

At this point we will have a total of (1,2,3,4,7) x 3= 51 lost.
In the recovery segment (9+10+11+12+13) =55, while maintaining the ratio of three to one, we will have a win of 4 tokens and a win of the previous winning shots (5,6,8) =19.

Conclusions

"Mater artium necessitas"
We conclude this book by taking up the Latin phrase with which it began, stressing that its content is only a very small part compared to who knows how many and which possible.
Sometimes just changing point of observation opens new possibilities and their applications can be useful in other fields.
So adding a book to the many others already out there, I want to publish my considerations on roulette, mathematical balances and bets, adding a new progression: The Venetian.

An evening at roulette

Monte Carlo Casino - Wednesday, December 24, 1930

N	R	N	R	N	R	N	R	N	R
	14	22			19		12	33	
24		35		24			32		1
	23		23		34	33			15
15			12	28		13		31	
	34	28		29		0	0	17	
	14	0	0	31		6			34
	14		14		3		14		1
	14	2		35			27		27
	7	17			24	15		35	
24			23		19	6		0	0
17			36		18	11			21
	32	10			9	22			18
	34	4			19	28			27
	14	28			32		23		25
33			12		32	35			34
4		29			1	28			12
29		26		20			18	4	
	25	17		15			27	28	
17		28			16	22			14
26			3		36		19	26	
	7	11			36	17			23
29		35		33		29			32
	34	24			22	6		6	
24			5	0	0	33			7

Martingale

N	R	bet	win	balance	N	R	bet	win	balance
	14	1	1	+1	22		2		+10
24		1		0	35		4		+6
	23	2	2	+2		23	8	8	+14
15		1		+1		12	1	1	+15
	34	2	2	+3	28		1		+14
	14	1	1	+4	0	0	2		
	14	1	1	+5		14	2		
	14	1	1	+6	2		2		+12
	7	1	1	+7	17		4		+8
24		1		+6		23	8	8	+16
17		2		+4		36	1	1	+17
	32	4	4	+8	10		1		+16
	34	1	1	+9	4		2		+14
	14	1	1	+10	28		4		+10
33		1		+9		12	8	8	+18
4		2		+7	29		1		+17
29		4		+3	26		2		+15
	25	8	8	+11	17		4		+11
17		1		+10	28		8		+3
26		2		+8		3	16	16	+19
	7	4	4	+12	11		1		+18
29		1		+11	35		2		+16
	34	2	2	+13	24		4		+12
24		1		+12		5	8	8	+20

N	R	bet	win	balance	N	R	bet	win	balance
	19	1	1	+21		12	4		
24		1		+20		32	4	4	+34
	34	2	2	+22	33		1		+33
28		1		+21	13		2		+31
29		2		+19	0	0	4		
31		4		+15	6		4		+27
	3	8	8	+23		14	8	8	+35
35		1		+22		27	1	1	+36
24		2		+20	15		1		+35
	19	4	4	+24	6		2		+33
	18	1	1	+25	11		4		+29
	9	1	1	+26	22		8		+21
	19	1	1	+27	28		16		+5
	32	1	1	+28		23	32	32	+37
	32	1	1	+29	35		1		+36
	1	1	1	+30	28		2		+34
20		1		+29		18	4	4	+38
15		2		+27		27	1	1	+39
	16	4	4	+31	22		1		+38
	36	1	1	+32		19	2	2	+40
	36	1	1	+33	17		1		+39
33		1		+32	29		2		+37
22		2		+30	6		4		+33
0	0	4			33		8		+25

N	R	bet	win	balance
33		16		+9
	1	32	32	+41
15		1		+40
31		2		+38
17		4		+34
	34	8	8	+42
	1	1	1	+43
	27	1	1	+44
35		1		+43
0	0	2		
	21	2		
	18	2	2	+45
	27	1	1	+46
	25	1	1	+47
	34	1	1	+48
	12	1	1	+49
4		1		+48
28		2		+46
	14	4	4	+50
26		1		+49
	23	2	2	+51
	32	1	1	+52
6		1		+51
	7	2	2	+53

D'Alembert

N	R	bet	win	balance	N	R	bet	win	balance
	14	1	1	+1	22		5		-2
24		1		0	35		6		-8
	23	2	2	+2		23	7	7	-1
15		1		+1		12	6	6	+5
	34	2	2	+3	28		5		0
	14	1	1	+4	0	0	6		
	14	1	1	+5		14	6		
	14	1	1	+6	2		6		-6
	7	1	1	+7	17		7		-13
24		1		+6		23	8	8	-5
17		2		+4		36	7	7	+2
	32	3	3	+7	10		6		-4
	34	2	2	+9	4		7		-11
	14	1	1	+10	28		8		-19
33		1		+9		12	9	9	-10
4		2		+7	29		8		-18
29		3		+4	26		9		-27
	25	4	4	+8	17		10		-37
17		3		+5	28		11		-48
26		4		+1		3	12	12	-36
	7	5	5	+6	11		11		-47
29		4		+2	35		12		-59
	34	5	5	+7	24		13		-72
24		4		+3		5	14	14	-58

N	R	bet	win	balance	N	R	bet	win	balance
	19	13	13	-45		12	10		
24		12		-57		32	10	10	-2
	34	13	13	-44	33		9		-11
28		12		-56	13		10		-21
29		13		-69	0	0	11		
31		14		-83	6		11		-32
	3	15	15	-68		14	12	12	-20
35		14		-82		27	11	11	-9
24		15		-97	15		10		-19
	19	16	16	-81	6		11		-30
	18	15	15	-66	11		12		-42
	9	14	14	-52	22		13		-55
	19	13	13	-39	28		14		-69
	32	12	12	-27		23	15	15	-54
	32	11	11	-16	35		14		-68
	1	10	10	-6	28		15		-83
20		9		-15		18	16	16	-67
15		10		-25		27	15	15	-52
	16	11	11	-14	22		14		-66
	36	10	10	-4		19	15	15	-51
	36	9	9	+5	17		14		-65
33		8		-3	29		15		-80
22		9		-12	6		16		-96
0	0	10			33		17		-113

N	R	bet	win	balance
33		18		-131
	1	19	19	-112
15		18		-130
31		19		-149
17		20		-169
	34	21	21	-148
	1	20	20	-128
	27	19	19	-109
35		18		-127
0	0	19		
	21	19		
	18	19	19	-108
	27	18	18	-90
	25	17	17	-73
	34	16	16	-57
	12	15	15	-42
4		14		-56
28		15		-71
	14	16	16	-55
26		15		-70
	23	16	16	-54
	32	15	15	-39
6		14		-53
	7	15	15	-38

Counter D'Alembert

N	R	bet	win	Balance	N	R	bet	win	balance
	14	1	1	+1	22		3		-11
24		2		-1	35		2		-13
	23	1	1	0		23	1	1	-12
15		2		-2		12	2	2	-10
	34	1	1	-1	28		3		-13
	14	2	2	+1	0	0	2		
	14	3	3	+4		14	2		
	14	4	4	+8	2		2		-15
	7	5	5	+13	17		1		-16
24		6		+7		23	1	1	-15
17		5		+2		36	2	2	-13
	32	4	4	+6	10		3		-16
	34	5	5	+11	4		2		-18
	14	6	6	+17	28		1		-19
33		7		+10		12	1	1	-18
4		6		+4	29		2		-20
29		5		-1	26		1		-21
	25	4	4	+3	17		1		-22
17		5		-2	28		1		-23
26		4		-6		3	1	1	-22
	7	3	3	-3	11		2		-24
29		4		-7	35		1		-25
	34	3	3	-4	24		1		-26
24		4		-8		5	1	1	-25

N	R	bet	win	balance	N	R	bet	win	balance
	19	2	2	-23		12	7		
24		3		-26		32	7	7	-8
	34	2	2	-24	33		8		-16
28		3		-27	13		7		-23
29		2		-29	0	0	6		
31		1		-30	6		6		-29
	3	1	1	-29		14	5	5	-24
35		2		-31		27	6	6	-18
24		1		-32	15		7		-25
	19	1	1	-31	6		6		-31
	18	2	2	-29	11		5		-36
	9	3	3	-26	22		4		-40
	19	4	4	-22	28		3		-43
	32	5	5	-17		23	2	2	-41
	32	6	6	-11	35		3		-44
	1	7	7	-4	28		2		-46
20		8		-12		18	1	1	-45
15		7		-19		27	2	2	-43
	16	6	6	-13	22		3		-46
	36	7	7	-6		19	2	2	-44
	36	8	8	+2	17		3		-47
33		9		-7	29		2		-49
22		8		-15	6		1		-50
0	0	7			33		1		-51

N	R	bet	win	balance
33		1		-52
	1	1	1	-51
15		2		-53
31		1		-54
17		1		-55
	34	1	1	-54
	1	2	2	-52
	27	3	3	-49
35		4		-53
0	0	3		
	21	3		
	18	3	3	-50
	27	4	4	-46
	25	5	5	-41
	34	6	6	-35
	12	7	7	-28
4		8		-36
28		7		-43
	14	6	6	-37
26		7		-44
	23	6	6	-38
	32	7	7	-31
6		8		-39
	7	7	7	-32

Limited counter D'Alembert

N	R	bet	win	balance	N	R	bet	win	balance
	14	1	1	+1	22		1		-8
24		1		-1	35		1		-9
	23	1	1	0		23	1	1	-8
15		2		-2		12	2	2	-6
	34	1	1	-1	28		3		-9
	14	2	2	+1	0	0	2		
	14	1	1	+1		14	2		
	14	1	1	+1	2		2		-11
	7	1	1	+1	17		1		-12
24		1		-1		23	1	1	-11
17		1		-2		36	2	2	-9
	32	1	1	-1	10		3		-12
	34	2	2	+1	4		2		-14
	14	1	1	+1	28		1		-15
33		1		-1		12	1	1	-14
4		1		-2	29		2		-16
29		1		-3	26		1		-17
	25	1	1	-2	17		1		-18
17		2		-4	28		1		-19
26		1		-5		3	1	1	-18
	7	1	1	-4	11		2		-20
29		2		-6	35		1		-21
	34	1	1	-5	24		1		-22
24		2		-7		5	1	1	-21

N	R	bet	win	balance	N	R	bet	win	balance
	19	2	2	-19		12	1		
24		3		-22		32	1	1	-1
	34	2	2	-20	33		2		-3
28		3		-23	13		1		-4
29		2		-25	0	0	1		
31		1		-26	6		1		-5
	3	1	1	-25		14	1	1	-4
35		2		-27		27	2	2	-2
24		1		-28	15		3		-5
	19	1	1	-27	6		2		-7
	18	2	2	-25	11		1		-8
	9	3	3	-22	22		1		-9
	19	4	4	-18	28		1		-10
	32	5	5	-13		23	1	1	-9
	32	6	6	-7	35		2		-11
	1	7	7	0	28		1		-12
20		8		-8		18	1	1	-11
15		7		-15		27	2	2	-9
	16	6	6	-9	22		3		-12
	36	7	7	-2		19	2	2	-10
	36	8	8	+6	17		3		-13
33		1		-1	29		2		-15
22		1		-2	6		1		-16
0	0	1			33		1		-17

N	R	bet	win	balance
33		1		-18
	1	1	1	-17
15		2		-19
31		1		-20
17		1		-21
	34	1	1	-20
	1	2	2	-18
	27	3	3	-15
35		4		-19
0	0	3		
	21	3		
	18	3	3	-16
	27	4	4	-12
	25	5	5	-7
	34	6	6	-1
	12	7	7	+6
4		1		-1
28		1		-2
	14	1	1	-1
26		2		-3
	23	1	1	-2
	32	2	2	0
6		3		-3
	7	2	2	-1

Dutch

N	R	bet	win	balance	sequence
	14	1	1	+1	
24		1		0	1
	23	2	2	+2	
15		1		+1	1
	34	2	2	+3	
	14	1	1	+4	
	14	1	1	+5	
	14	1	1	+6	
	7	1	1	+7	
24		1		+6	1
17		2		+4	12
	32	2	2	+6	2
	34	3	3	+9	
	14	1	1	+10	
33		1		+9	1
4		2		+7	12
29		2		+5	122
	25	2	2	+7	22
17		3		+4	223
26		3		+1	2233
	7	3	3	+4	233
29		3		+1	2333
	34	3	3	+4	333
24		4		0	3334

N	R	bet	win	balance	sequence
22		4		-4	33344
35		4		-8	333444
	23	4	4	-4	33444
	12	4	4	0	3444
28		4		-4	34444
0	0	4			
	14	4			
2		4		-8	344444
17		4		-12	3444444
	23	4	4	-8	444444
	36	5	5	-3	44444
10		5		-8	444445
4		5		-13	4444455
28		5		-18	44444555
	12	5	5	-13	4444555
29		5		-18	44445555
26		5		-23	444455555
17		5		-28	4444555555
28		5		-33	44445555555
	3	5	5	-28	4445555555
11		5		-33	44455555555
35		5		-38	444555555555
24		5		-43	4445555555555
	5	5	5	-38	445555555555

47

N	R	bet	win	balance	sequence
	19	5	5	-33	45555555555
24		5		-38	455555555555
	34	5	5	-33	55555555555
28		6		-39	555555555556
29		6		-45	5555555555566
31		6		-51	55555555555666
	3	6	6	-45	5555555555666
35		6		-51	55555555556666
24		6		-57	555555555566666
	19	6	6	-51	55555555566666
	18	6	6	-45	5555555566666
	9	6	6	-39	555555566666
	19	6	6	-33	55555566666
	32	6	6	-27	5555566666
	32	6	6	-21	555566666
	1	6	6	-15	55566666
20		6		-21	555666666
15		6		-27	5556666666
	16	6	6	-21	556666666
	36	6	6	-15	56666666
	36	6	6	-9	6666666
33		7		-16	66666667
22		7		-23	666666677
0	0	7			

N	R	bet	win	balance	sequence
	12	7			
	32	7	7	-16	66666677
33		7		-23	666666777
13		7		-30	6666666777
0	0	7			
6		7		-37	66666677777
	14	7	7	-30	6666677777
	27	7	7	-23	666677777
15		7		-30	6666777777
6		7		-37	66667777777
11		7		-44	666677777777
22		7		-51	6666777777777
28		7		-58	66667777777777
	23	7	7	-51	6667777777777
35		7		-58	66677777777777
28		7		-65	666777777777777
	18	7	7	-58	66777777777777
	27	7	7	-51	6777777777777
22		7		-58	67777777777777
	19	7	7	-51	7777777777777
17		8		-59	77777777777778
29		8		-67	777777777777788
6		8		-75	7777777777777888
33		8		-83	77777777777778888

N	R	bet	win	balance	sequence
33		8		-91	7777777777777788888
	1	8	8	-83	777777777777788888
15		8		-91	777777777777888888
31		8		-99	7777777777778888888
17		8		-107	77777777777788888888
	34	8	8	-99	7777777777788888888
	1	8	8	-91	777777777788888888
	27	8	8	-83	77777777788888888
35		8		-91	777777777888888888
0	0	8			
	21	8			
	18	8	8	-83	77777777888888888
	27	8	8	-75	7777777888888888
	25	8	8	-68	777777888888888
	34	8	8	-60	77777888888888
	12	8	8	-52	7777888888888
4		8		-60	77778888888888
28		8		-68	777788888888888
	14	8	8	-60	77788888888888
26		8		-68	777888888888888
	23	8	8	-60	77888888888888
	32	8	8	-52	7888888888888
6		8		-60	78888888888888
	7	8	8	-52	8888888888888

The Venetian

N	R	bet	win	balance	N	R	bet	win	balance
	14	1	1	+1	22		3		-2
24		1		0	35		3		-5
	23	1	1	+1		23	4	4	-1
15		1		0		12	4	4	+3
	34	1	1	+1	28		4		-1
	14	1	1	+2	0	0	4		
	14	1	1	+3		14	4		
	14	1	1	+4	2		4		-5
	7	1	1	+5	17		4		-9
24		1		+4		23	4	4	-5
17		1		+3		36	4	4	-1
	32	2	2	+5	10		4		-5
	34	1	1	+6	4		4		-9
	14	1	1	+7	28		4		-13
33		1		+6		12	5	5	-8
4		1		+5	29		5		-13
29		2		+3	26		5		-18
	25	2	2	+5	17		5		-23
17		2		+3	28		6		-29
26		2		+1	3		6	6	-23
	7	3	3	+4	11		6		-29
29		3		+1	35		6		-35
	34	3	3	+4	24		7		-42
24		3		+1		5	7	7	-35

N	R	bet	win	balance	N	R	bet	win	balance
	19	7	7	-28		12	2		
24		7		-35		32	2	2	+7
	34	7	7	-28	33		1		+6
28		7		-35	13		1		+5
29		7		-42	0	0	2		
31		7		-49	6		2		+3
	3	8	8	-41		14	2	2	+5
35		8		-49		27	2	2	+7
24		8		-57	15		1		+6
	19	8	8	-49	6		1		+5
	18	8	8	-41	11		2		+3
	9	8	8	-33	22		2		+1
	19	8	8	-25	28		3		-2
	32	8	8	-17		23	3	3	+1
	32	8	8	-9	35		3		-2
	1	8	8	-1	28		3		-5
20		8	8	-9		18	4	4	-1
15		8	8	-17		27	4	4	+3
	16	8	8	-9	22		4		-1
	36	8	8	-1		19	4	4	+3
	36	8	8	+7	17		4		-1
33		1		+6	29		4		-5
22		1		+5	6		4		-9
0	0	2			33		4		-13

N	R	bet	win	balance
33		5		-18
	1	5	5	-13
15		5		-18
31		5		-23
17		6		-29
	34	6	6	-23
	1	6	6	-17
	27	6	6	-11
35		6		-17
0	0	6		
	21	6		
	18	6	6	-11
	27	6	6	-5
	25	6	6	+1
	34	6	6	+7
	12	1	1	+8
4		1		+7
28		1		+6
	14	2	2	+8
26		1		+7
	23	1	1	+8
	32	1	1	+9
6		1		+8
	7	1	1	+9

Progressions

N	R	Martingale	D'Alembert	Counter	Limited	Dutch	Venetian
	14	+1	+1	+1	+1	+1	+1
24		0	0	-1	-1	0	0
	23	+2	+2	0	0	+2	+1
15		+1	+1	-2	-2	+1	0
	34	+3	+3	-1	-1	+3	+1
	14	+4	+4	+1	+1	+4	+2
	14	+5	+5	+4	+1	+5	+3
	14	+6	+6	+8	+1	+6	+4
	7	+7	+7	+13	+1	+7	+5
24		+6	+6	+7	-1	+6	+4
17		+4	+4	+2	-2	+4	+3
	32	+8	+7	+6	-1	+6	+5
	34	+9	+9	+11	+1	+9	+6
	14	+10	+10	+17	+1	+10	+7
33		+9	+9	+10	-1	+9	+6
4		+7	+7	+4	-2	+7	+5
29		+3	+4	-1	-3	+5	+3
	25	+11	+8	+3	-2	+7	+5
17		+10	+5	-2	-4	+4	+3
26		+8	+1	-6	-5	+1	+1
	7	+12	+6	-3	-4	+4	+4
29		+11	+2	-7	-6	+1	+1
	34	+13	+7	-4	-5	+4	+4
24		+12	+3	-8	-7	0	+1

N	R	Martingale	D'Alembert	Counter	Limited	Dutch	Venetian
22		+10	-2	-11	-8	-4	-2
35		+6	-8	-13	-9	-8	-5
	23	+14	-1	-12	-8	-4	-1
	12	+15	+5	-10	-6	0	+3
28		+14	0	-13	-9	-4	-1
0	0						
	14						
2		+12	-6	-15	-11	-8	-5
17		+8	-13	-16	-12	-12	-9
	23	+16	-5	-15	-11	-8	-5
	36	+17	+2	-13	-9	-3	-1
10		+16	-4	-16	-12	-8	-5
4		+14	-11	-18	-14	-13	-9
28		+10	-19	-19	-15	-18	-13
	12	+18	-10	-18	-14	-13	-8
29		+17	-18	-20	-16	-18	-13
26		+15	-27	-21	-17	-23	-18
17		+11	-37	-22	-18	-28	-23
28		+3	-48	-23	-19	-33	-29
	3	+19	-36	-22	-18	-28	-23
11		+18	-47	-24	-20	-33	-29
35		+16	-59	-25	-21	-38	-35
24		+12	-72	-26	-22	-43	-42
	5	+20	-58	-25	-21	-38	-35

N	R	Martingale	D'Alembert	Counter	Limited	Dutch	Venetian
	19	+21	-45	-23	-19	-33	-28
24		+20	-57	-26	-22	-38	-35
	34	+22	-44	-24	-20	-33	-28
28		+21	-56	-27	-23	-39	-35
29		+19	-69	-29	-25	-45	-42
31		+15	-83	-30	-26	-51	-49
	3	+23	-68	-29	-25	-45	-41
35		+22	-82	-31	-27	-51	-49
24		+20	-97	-32	-28	-57	-57
	19	+24	-81	-31	-27	-51	-49
	18	+25	-66	-29	-25	-45	-41
	9	+26	-52	-26	-22	-39	-33
	19	+27	-39	-22	-18	-33	-25
	32	+28	-27	-17	-13	-27	-17
	32	+29	-16	-11	-7	-21	-9
	1	+30	-6	-4	0	-15	-1
20		+29	-15	-12	-8	-21	-9
15		+27	-25	-19	-15	-27	-17
	16	+31	-14	-13	-9	-21	-9
	36	+32	-4	-6	-2	-15	-1
	36	+33	+5	+2	+6	-9	+7
33		+32	-3	-7	-1	-16	+6
22		+30	-12	-15	-2	-23	+5
0	0						

N	R	Martingale	D'Alembert	Counter	Limited	Dutch	Venetian
	12						
	32	+34	-2	-8	-1	-16	+7
33		+33	-11	-16	-3	-23	+6
13		+31	-21	-23	-4	-30	+5
0	0						
6		+27	-32	-29	-5	-37	+3
	14	+35	-20	-24	-4	-30	+5
	27	+36	-9	-18	-2	-23	+7
15		+35	-19	-25	-5	-30	+6
6		+33	-30	-31	-7	-37	+5
11		+29	-42	-36	-8	-44	+3
22		+21	-55	-40	-9	-51	+1
28		+5	-69	-43	-10	-58	-2
	23	+37	-54	-41	-9	-51	+1
35		+36	-68	-44	-11	-58	-2
28		+34	-83	-46	-12	-65	-5
	18	+38	-67	-45	-11	-58	-1
	27	+39	-52	-43	-9	-51	+3
22		+38	-66	-46	-12	-58	-1
	19	+40	-51	-44	-10	-51	+3
17		+39	-65	-47	-13	-59	-1
29		+37	-80	-49	-15	-67	-5
6		+33	-96	-50	-16	-75	-9
33		+25	-113	-51	-17	-83	-13

N	R	Martingale	D'Alembert	Counter	Limited	Dutch	Venetian
33		+9	-131	-52	-18	-91	-18
	1	+41	-112	-51	-17	-83	-13
15		+40	-130	-53	-19	-91	-18
31		+38	-149	-54	-20	-99	-23
17		+34	-169	-55	-21	-107	-29
	34	+42	-148	-54	-20	-99	-23
	1	+43	-128	-52	-18	-91	-17
	27	+44	-109	-49	-15	-83	-11
35		+43	-127	-53	-19	-91	-17
0	0						
	21						
	18	+45	-108	-50	-16	-83	-11
	27	+46	-90	-46	-12	-75	-5
	25	+47	-73	-41	-7	-68	+1
	34	+48	-57	-35	-1	-60	+7
	12	+49	-42	-28	+6	-52	+8
4		+48	-56	-36	-1	-60	+7
28		+46	-71	-43	-2	-68	+6
	14	+50	-55	-37	-1	-60	+8
26		+49	-70	-44	-3	-68	+7
	23	+51	-54	-38	-2	-60	+8
	32	+52	-39	-31	0	-52	+9
6		+51	-53	-39	-3	-60	+8
	7	+53	-38	-32	-1	-52	+9

www.ingramcontent.com/pod-product-compliance
Lightning Source LLC
Chambersburg PA
CBHW070851220526
45466CB00005B/1960